ECCRINE POWER

Giacomo Fasano

Author's Tranquility Press
ATLANTA, GEORGIA

Copyright © 2024 by Giacomo Fasano

All rights reserved. No part of this publication may be reproduced, distributed or transmitted in any form or by any means, including photocopying, recording, or other electronic or mechanical methods, without the prior written permission of the publisher, except in the case of brief quotations embodied in critical reviews and certain other noncommercial uses permitted by copyright law. For permission requests, write to the publisher, addressed "Attention: Permissions Coordinator," at the address below.

Giacomo Fasano/Author's Tranquility Press

3900 N Commerce Dr. Suite 300 #1255

Atlanta, GA 30344, USA

www.authorstranquilitypress.com

Ordering Information:

Quantity sales. Special discounts are available on quantity purchases by corporations, associations, and others. For details, contact the "Special Sales Department" at the address above.

Eccrine Power / Giacomo Fasano
Hardback: 978-1-964362-66-3
Paperback: 978-1-964362-13-7
eBook: 978-1-964362-14-4

Contents

Preamble..

Conclusion...

Preamble Conclusion ....

What is Eccrine Power? 8

HOW SCIENCE GOT IT WRONG AND WHAT THEY
GOT RIGHT! 9

Glossary of Scientific Terms and Facts 30

    The Role of Lymphatic Capillaries 34

Glossary 36

Sample Diets: 38

"Run & Sun" 44

Eccrine "Star" Power..47

Concluding Quote 73

Preamble

"A Preamble is an introductory statement that provides context or background information about a document or media piece."

The purpose of Eccrine Power is to reduce illnesses and prolong lives worldwide! Along with this comes increasing one's self-worth and happiness, through optimal health! Of course, to correct the science dictionaries on the purpose of Eccrine glands. If this generation of scientists and human beings do not want to fully adopt it and accredit its findings, so be it! It's fine. Eccrine Power is built to stand the test of time! It is truly timeless data built purely of facts and real science. It will never get old and eventually, whether its 50, 100, 200 or 500 years or greater, a future generation will go crazy & viral over it and bring back life-spans of those only found in biblical times of 200, 350, and even 950 years old! Eventually, all the chips will fall in place and its utilization will be mastered and growth in longevity achieved! The only question truly is? Which generation will be the smartest and the fastest?

Conclusion

(So compelling, it needs to go in the beginning!)

Getting this Eccrine gland Science correct is a pretty big deal actually... This, since currently it's Scientifically defined as nonchalant; involuntary action that's incorrect conclusion of purpose is "to cool the body down when hot", whereas mine is a voluntary action that's critical to your life health and purification. So yeah, I think it needs to be addressed and that it is a rather big piece of "Missing Science"!! Let me tell you how big!

Science currently knows EVERYTHING there is to know about the actual composition of the human body and how it works... No, they really do. EVERYTHING! Except a simple oversight and wrong conclusion of the analysis of the base section of the Eccrine gland. Specifically, pointing all its function to the nerve network as opposed to the blood network (blood capillary lines)!! It's as simple as that! However, the impacts are grossly far-reaching and deadlier than anyone could have ever imagined. An oversight/mistake so clearly obvious in its facts and visual presentation; its downright SCARY!!

YOUR body is a massive pool of water and we all know a pool needs a water purification process. Remember, this process cleans the inside of the body which you cannot see! Hmmm, here it is... Eccrine Power and what is in it for you? Well, if you begin to use the magic of "E-Power (Eccrine Power) just 3-6 hours a week by running these Eccrine glands, you can:

1. Turn back your appearance up 20 years and more importantly, clean the insides of your body, which again, you cannot see. (Ok, if not 20 years how about a few and keeping where you are now!!)

2. Lower your risk of Heart attack, stroke, diabetes, cancer, alcoholism,

and depression GUARANTEED by removing some of the factually proven root causes such as heavy metals, fats, sugars, salts, alcohol, and micro-plastics DIRECTLY from your inside!! Specifically, your CELLS & BLOOD!! Oh well, it seems Science won't and can't tell you this since they want you to believe the ECCRINE PORES run of the Nerve network... hahaha, that is so funny and comical, its ridiculous! Talk about "Junk Science" here it is... I SWEAT 3-5 liters of water-based sweat from my insides and it is controlled by the nerve network not the blood lines that go to the entire blood flowing body... yeah ok, have another shot of whiskey!! DUH! How lame... how stupid, and how wrong!! So only you and I can change this mishap by Science. AND WE WILL!!

3. Sleep better, eat better, think clearer, remember more, feel less stress, have more energy, feel more confident and AGAIN look better. Look better = feel better!

4. Feel Accomplishment- finishing 68 minutes a session of E-Power will gain you purpose and sense of achievement.

5. Conform to your bodies core purpose and design by God. Utilize the theory that God designed our bodies with specific purpose and as I've said' We are built to Run, Built to Sweat, and Built to Self-Heal! This is the "ONLY Way to feel and get younger as you get older" An oxy moron till now. STAY YOUNG FOREVER!!

6. 6. SPIRITUAL GROWTH – I've feel the power of 18+ years of doing this vital-cycle! You will grow in faith and spiritual confidence simply by your focused meditation and directed prayer to your God as you perform the various cardio activity necessary to push through an hour straight. IT ONLY GETS BETTER WITH TIME AND YEARS!

Preamble Conclusion

Eccrine Power or E-power as I like to call it; creates a duty to every person on the planet towards their own body and its large water-base!

It makes it incumbent upon themselves to recognize the 800 lb gorilla in their body! That gorilla is in fact, their 42 L of water. You wouldn't ignore a gorilla, so they shouldn't ignore this water. It's equivalent of 42 L of water in one captured surrounding (body). The duty is simple, move your God-given limbs vigorously periodically to experience the magic flush of Eccrine power!

What is Eccrine Power?

Eccrine power is the use of over 400,000 tiny microscopic hoses stowed away hidden in your skin that are microscopically attached to blood lines called blood capillaries that run directly to all your inside organs to manage pressure from increased heart rate.

This, ultimately means the purification of this water since it's built on pressure! Two ways to raise heart rate. First, heat causes heat stress and increases heart rate which can make you sweat from the Eccrine pores. More natural is motion which gradually causes the heart rate to increase with effort. This means the more heart rate, the more resulting pressure from the faster flow of blood to the organs. This translates into a "Scientific" term known as hydro-static pressure. This pressure is transferred to the same Eccrine pores through blood capillary connections on both ends.

None of this has been recognized! since as a diagram shows science somehow focused entirely on the nerve network as the base element of these most effective and powerful glands... this is a mind-boggling oversight that simply has no basis on how it is currently defined. The nerve network has no connections to water in the body whatsoever. If anything, the nerve network can get your heart rate to go up when you're nervous and produce sweat through again, an increased heart rate. It all comes down to increased heart rate! Everything operates from that on these glands.

Nerves + no increased heart rate = No Sweat produced!

Nerves + increased heart rate = Sweat produced

No Nerves + increased heart rate (from movement or heat) = Sweat Produced!

The higher, the heart rate = more sweat emitted from Eccrine pores!

So, the harder you run or climb stairs and the longer, the higher your heart rate and resulting pressure and the MORE SWEAT OUT THE PORES!

THE MORE HEAT AND HIGHER THE HEAT STRESS, WITH OR WITHOUT BODILY MOVEMENT, THE MORE HEART RATE INCREASE AND HENCE SAME RELEASE FROM PORES. WHAT IS MORE NATURAL TO RAISE YOUR HEART RATE AND SWEAT ECCRINE POWER? HEAT OR MOVEMENT??

ANSWER- IS NOT A TRICK QUESTION, GODS WAY IS MOVEMENT TO SWEAT OUT THE PORES. HEAT ONLY WITH NO MOVEMENT HAS RISKS AFTER ONLY 12-20 MINUTES SINCE YOUR HEART RATE IS ELEVATING TO DEAL WITH THE HEAT CAUSING STRESS AND NO BODY MOVEMENT... THIS IS WHY IT'S IMPORTANT TO CHOOSE MOVEMENT NOT HEAT (SAUNA, ETC.) STROKES AND HEART ATTACKS MAINLY ONLY OCCUR FROM HEAT WITH NO MOVEMENT DUE TO THIS FACTOR. UNDERSTAND? END OF THE DAY, RUN OR EQUIVALENT, DON'T JUST JUMP IN THE SAUNA.

This is undeniable science!

There's nothing quite in the world that can do what Eccrine power can do and that is make your body like fine wine and improve with time !!! by factually extracting all the bad in the pool of water of your body as opposed to the opposite, fermenting it and let destroy your inside and outside and its natural course of rapid Aging!!

A common-sense approach to studying ANY "Pool of water" makes this study of a pool of water that's in your own body— UNDENIABLE because any "Pool off water" has to have a filtration process... this is common sense undeniable and applies to filtration of the human pool of water which you can't live without and it is what it is. Eccrine power IS YOUR BODIES WATER POOL'S FILTRATION SYSTEM FOR ALL YOUR INSIDE CELLS!!

I always like to start a book off with a Quiz/Test, in this case one simple question to draw your attention properly...

Question # 1

How would you best explain the way medicines enter your body from a "Medical (transdermal) Patch" mainly used in modern days on any one of the arms by attaching it to the surface?

Answer-

The medicines enter the body through these very Eccrine Pores and their blood capillary lines as the highway to all internal blood flow, in reverse fashion of "Eccrine Power" which simply uses the Outflow aspect of them as opposed to inflow...

"The funny part is that I don't even know if the manufacturers even know this since they just only say it enters the bloodstream through the skin..."

This books finding could be the most important aspect of our water-based body... yet it has been completely ignored? How can that be? One would have to ask themselves the question. Is it because the fact that our bodies water-base AND THAT being the most important aspect TO LONG TERM HEALTH AND VITALITY has not been recognized, or given the attention, it deserves? I think so and quite frankly, it's because of the lack of discovery and abstract facts that I have presented regarding the largest carries of water in the body and how the body is all built around the water. **Through these never theorized blood capillary lines on both ends of the internal body, specifically the tying of the blood capillary lines from the Eccrine pores DIRECTLY TO A VITAL ORGAN and its BLOOD CAPILLARY LINES ON THE SURFACES of the tissues of those organs**

-Giacomo Fasano

You would think this has already been prioritized, but in fact, it has not! Till now!

Welcome to Eccrine Power and welcome to your new beginning!

Scientists, geniuses, scholars, laymen, Masters of trade, all will confer with these findings because they are 100% undeniable based on facts, common sense and logic which is all anybody needs to conclude on a matter! ***AGING IS NOT REVERSIBLE!! HOWEVER, ITS BASIS IS MAJORITY ABLE TO BE HALTED***, with this life saving process of bodily water purification! In the beginning, God made it all, and then he picked the water base of the human beings, and the creator made the creatures, or our God made us! Then he made us mobile, extremely mobile so we can move AND run about with our limbs that demonstrate that succinctly. Most importantly, he connected our whole bodies internal parts to outlets, called the Eccrine pores that would allow us to PRESSURE OUT the bad from inside of our body over time.

Just imagine this. You've purchased this book for a small investment and now in return. I am going to give you a tool to purify all your inner vital organs. We're talking about a very impressive list that is pictured below. They will be healthier, more vital and ready for a long-term life. That's healthy and filled with vitality. I want to do it by attaching blood vessels like suction cups to the tissues of every organ and I'm gonna run those lines directly through the skin to a pore and a hose and which will only react to the pressure that you can control with your movement and increase heart rate. Simple as that. In the following key Science diagram, you can easily see how each vital organ is made up of a high percentage of water, and that is begging for these attachments to be used to purify. If you don't do the Eccrine Powercycle you can simply imagine with common sense how any pool of water, regardless of its base or form (organ tissue & cells) will degrade over time when introduced to outside elements. Same rules apply to your internal sub-pools of water that are each one of your organs, vital or non-vital!

Eccrine Power

So, tell your family, tell your friends, tell your coworkers, you have bought a book that is teaching you how to purify all your water-based organs, like your brain, kidneys, liver, stomach, etc. all because the water-based organs have bloodlines connected to them and all because they already receive blood flow from the heart.

Want to learn more? Of course you do! Come along for the magical ride of your life! God made us! there is no other way and he deserves all the glory. He made us with precision and accuracy. He made us mobile and water-based to encompass the power of moving water to our complex, yet simple body. We need food, air, and water to live! Our bodies thrive around the trio. However, what it doesn't want is the toxic byproducts and trace remains of these big three in our cells and body parts long term (sugars, salts, heavy metals, fats, cholesterol, microplastics, etc... This is where the problem is inherent amongst most Americans and people worldwide. They are not using heavy cardio to blast out these detriments regularly like world class soccer players. They are not because they are NOT aware of this groundbreaking discovery and theory. The VITAL BLOOD CAPILLARY LINES THAT YOU WILL LEARN ARE CONNECTING TO YOUR ORGANS RUN DIRECTLY TO THE EXITPOINT OF THE HOUSE OF THE SWEAT ECCRINE GLANDS WHOSE BLOOD CAPILLARY LINE ATTACHMENTS HAVE ALSO NEVER BEEN IDENTIFIED WITH THEORY TO RUN DIRECTLY TO THE ORGANS AS RELIEF FROM HYDROSTATIC PRESSURE (faster heart rate and blood flow from heart to all organs). IT IS THEREFORE THAT THIS BOOK IS FIRST TO STAKE CLAIM THAT ALL PARTS OF THE INSIDE OF THE HUMAN BODY HAVE THESE BLOOD CAPILLARY LINES AND THAT THEY SOLE PURPOSE IS TO RECEIVE AND TRANSFER OUT CELLULAR WATER FROM THE CELLS FROM THE HYDROSTATIC PRESSURE AND ALONG WITH THIS

MOVING WATER DRAGS UNWANTED DETRIMENTS DESCRIBED ABOVE1

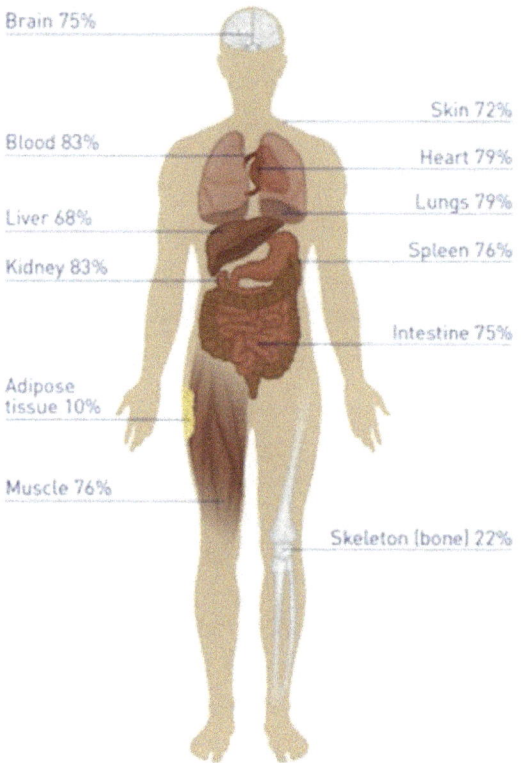

The above diagram shows the amazing high percentage of water that each organ is made up of. For example, when it states 75% pointing to the brain, it means the brain is made up of 75% WATER!! Every part of our insides down to even out bones have water!! And most a very high percent!! Yes, you want to know more about the blood capillaries running from them to the ECCRINE PORES!!

We should be happy in life! Very happy and thankful! For God has made us, and he has made us water-based for us to thrive overtime. Yes, we were created, water-based and crafted by intelligent design

Eccrine Power

by the most powerful almighty, and since only horses & humans have these glands with these complex blood connections; we couldn't have evolved from Apes either! That's right, I'll say it again: we could not have evolved from apes since they don't not have these complex connections or hundreds of thousands of eccrine pores, leaving only CREATION left since evolution just got knocked out. No really, this was discussed in my book prior, but quite simply, there is No way along the lines of the Sciences diagram of evolution that we could have just all of the sudden spawned these glands and blood connections which run from the organs to the pore! Since there are only really two theories predominant in the world today, Evolution and Creation. If I factually remove one in evolution we are only left with God's creation.

IT'S A SCIENTIFIC IMPOSSIBILITY FOR ANYWHERE ALONG THE DEPICTED LINEAGE OF APES TO HUMANS FOR THAT BODY TO HAVE MAGICALLY SPAWNED THE MILLIONS OF BLOOD CAPILLARIES RUNNING TO ECCRINE PORES FROM ORGAN TISSUE. IMPOSSIBLE!

END OF STORY

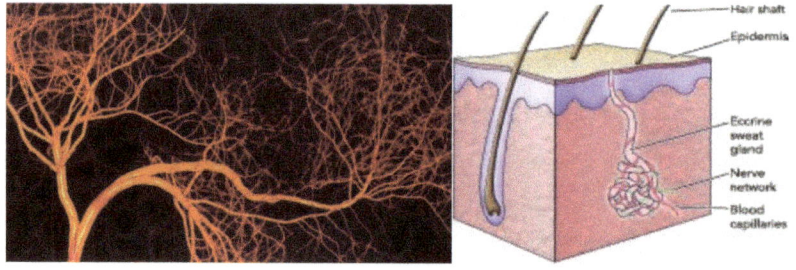

When I say we couldn't have evolved from apes, I mean we couldn't have evolved with 100% certainty!! Yes, 100% and when you take this away from Science, they really have no theory of evolution. It is God and God only who made our bodies, and by design. This fits perfectly into the equation as water is known to be the most powerful substance in the world, so why wouldn't a creator use it when making a human body such as ours when you see this perspective with the bloodlines connecting from organ tissue to open pores or hoses, you realize the magnitude of why we were built to run and sweat, I have seen what the world is doing and it is not at all. The time is now to change for the future generations. Eccrine power will do that by allowing younger and older folks to manipulate the water base and extract the bad repeatedly over time.

Then as I always say, we were Built to Run (limbs), Built to Sweat (pressure through movement through the pores), and BUILT to Self-Heal (through our purification of our internal water-base)

The proof is in the pudding: those who run get paid to run look at them. The bottom line is if you don't do this process YOU WILL LOSE TO THE Aging battle. I'm writing this book not only because of the enormity of the Eccrine glands blood lines to organs discovery, but also to try to share with the world how I, Giacomo Fasano, is stealing time on a daily basis, like a thief, taking time from Aging by manipulating these eccrine glands with a HIGH USAGE and EXIT

PLAN of water from my insides. THIS PROCESS, WITH THESE OVERLOOKED BLOOD CONNECTIONS IS THE KEY TO STAYING YOUNG FOREVER!

As I begin, it is most obvious that if we all simply acknowledge the simple fact that every human being on the planet is water-based and carries an average of 42 liters of water, making up **2/3** of the body! TO ME, IT SHOULD BE NO SURPRISE THAT WE WOULD NEED TO PROACTIVELY HARD PRESS A LITER OR TWO ALMOST DAILY FOR BASIC COMMON-SENSE CLEANLINESS OF OUR INSIDES. This is exactly what the JETSTREAM PowerCycle does! HARD Pressure water out of our body by pressure!! born from THE MOST POWERFUL TOOL ON THE PLANET...Just what tool might that be, you ask? Well not a bobcat since it's stronger than that, also not a crane since its power is much more succinct! The strongest tool to mankind is none other than our totally ignored JETSTREAM blood circulation! Yes, I said our Jet-stream is like blood circulation. Even with a normal heart rate of only 72 beats per minute (bpm's) it is traveling at LIGHT SPEEDS in and throughout our body!! It is undeniably a titanic force of active energy traveling very fast... This is a very important FACT! Since it is the most powerful tool to mankind, can anything be more powerful?? Yes, only one other thing can be. That is that same **JETSTREAM blood circulation** on steroids or with a turbo boost button... Or eliminate that slang and simply state THE ONLY THING MORE POWERFUL IS THIS SAME JETSTREAM BLOOD CIRCULATION when elevated higher in supersonic speeds with is none other than you raising your heart rate up to increase the pumps and beats per minute to create an internal thrust which Science already knows as HYDROSTATIC pressure, the direct result of this TOP power tool in action when forced by movement to increase its powerful flow to all that the HEART PUMPS BLOOD TOO! I come at you with the strongest line of them all. I have found a formula that can allow you as a person to become stronger, younger, and even better looking as you get older. This by stealing the good from all available sources of diet, including all foods, and even bad to strengthen the cells with all the unique nutrients and elements that are available in the entire world of diet and supplements, but with the key element of after strengthening those cells with that also kicking all the bad in ONE SIMPLE PROCESS! This is the force of the "JETSTREAM Powercycle" as part of the Powercycle ...

Again, imagine a process that allows you to strengthen all the cells of your body with all unique good and bad elements of life and then kicking out all the bad after that strengthening process in one symbol process

Next, I unveiled is powerful tool that's been overlooked, and they were discovered that all needed to be identified through connecting of the tree branches of blood capillary lines

So wait, STOP! Let's absorb this right now. You have bought this book and now learned you have your very own custom POWER TOOL called your JETSTREAM blood circulation. It is yours for unlimited use and you have likely ignored it till now...?

It is literally the tool to make your STAY YOUNG FOREVER and feel young all along. It can use its powers to thoroughly clean your interior cells and blood, which you will learn over and over are the TOP 1 and 2 carriers of water in your body. Most importantly you are going to learn that this process has yet to be Scientifically accredited because of a simple oversight of blood capillaries at the Eccrine gland outlets.

HOW SCIENCE GOT IT WRONG AND WHAT THEY GOT RIGHT!

Science has studied and gotten one very important FACT concluded which turns out to be the Front-end of the Powercycle equation and that is HYDROSTATIC PRESSURE. As I have stated many times throughout, they studied exercise and elevated heart rate and resulting blood pressure increase from and concluded that when a person raised their heart rate from moving that this pressure causes cellular water to transfer from our internal cells into our BLOODSTREAM! That is, it. NOT where it goes, just that this process occurs. Then where they got it wrong is when they didn't connect the Blood capillaries tree-like connections as the outlet for this process. Then since that was missed, they didn't do the simple thing that I have which is to connect the blood capillaries at the eccrine pores to the blood capillaries attaching to all tissues of our organs!

If they did, they would be the ones delivering this powerful news! That which is your JETSTREAM BLOOD CIRCULATION is the tool to cleanse out all our cells of unwanted toxins and detriments in a very planned out, and systematic process which was God given and designed.

So basic it is as well. The heart already pumps blood to the entire body. Revolutionary Powercycles and the JETSTREAM Powercycle is stating that any organ or internal tissue that receives blood from the heart (ALL Body), MUST, and I emphasize MUST also have blood capillaries running to an eccrine pore or it would not be able to deal with Sciences ``Hydro-static pressure from increased heart rate and blood circulation. Simple as that. This in essence makes these blood line connections from the organs to the eccrine pores a new part of the organ as I like to dub, "Octopus legs" as they similarly run from the head or base of the organ to the outlet yet are part of the organ since the water flowing from pressure from cells is directly traveling to its release point.

This is so fundamentally sound and basic in nature, yet completely ignored till now!

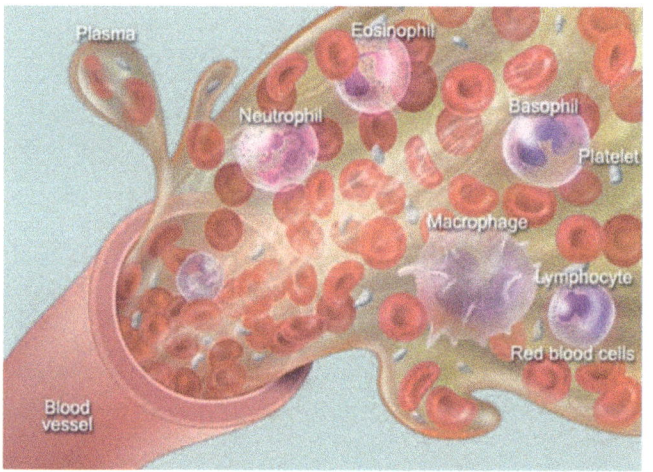

Pictured above – The **JETSTREAM NATURE OF OUR BLOOD CIRCULATION!** FAST and powerful!! Ready to be boosted and cause pressure cleaning and release of cellular water marred with trace toxins and detriments (Salts, sugars, fats, heavy metals).

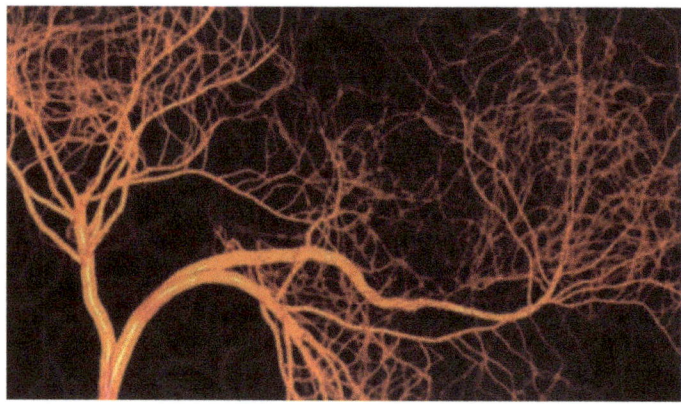

The unique "Tree like" Blood capillary lines pictured above are tied together from the Eccrine pore sack to the surfaces of all vital organs in the body with a DIRECT connection from the uniques pipeline that trees from one end (Eccrine Pores) to the other end (organ tissue surface) to release "Hydro-static pressure" from increased heart rate which can only be caused by two methods: 1) Movement (preferred and with no max time or unlimited)) and or 2) Heat (not preferred and max time 20 minutes)

REAL LIFE BLOOD CAPILLARIES FROM THE HUMAN BODY AND WHAT CONNECTS TO THE HUNDREDS OF THOUSANDS ECCRINE GLANDS (BOTTOM PHOTO above-2nd photo)

Human Skin Diagram
Hair
Stratum Corneum
Granular Cell Layer
Spinous Cell Layer
Basal Cell Layer
Sebaceous Gland
Erector Pili Muscle
Sweat Gland
Hair Follicle
Collagen And
Elastin Fibres
Artery
Vein
Hair shaft
Epidermis
Eccrine
sweat
gland
Nerve
network
Blood
capillaries

Plasma
Eosinophil
Neutrophil
Basophil
Platelet
Macrophage
Lymphocyte
Red blood cells
Blood
vessel

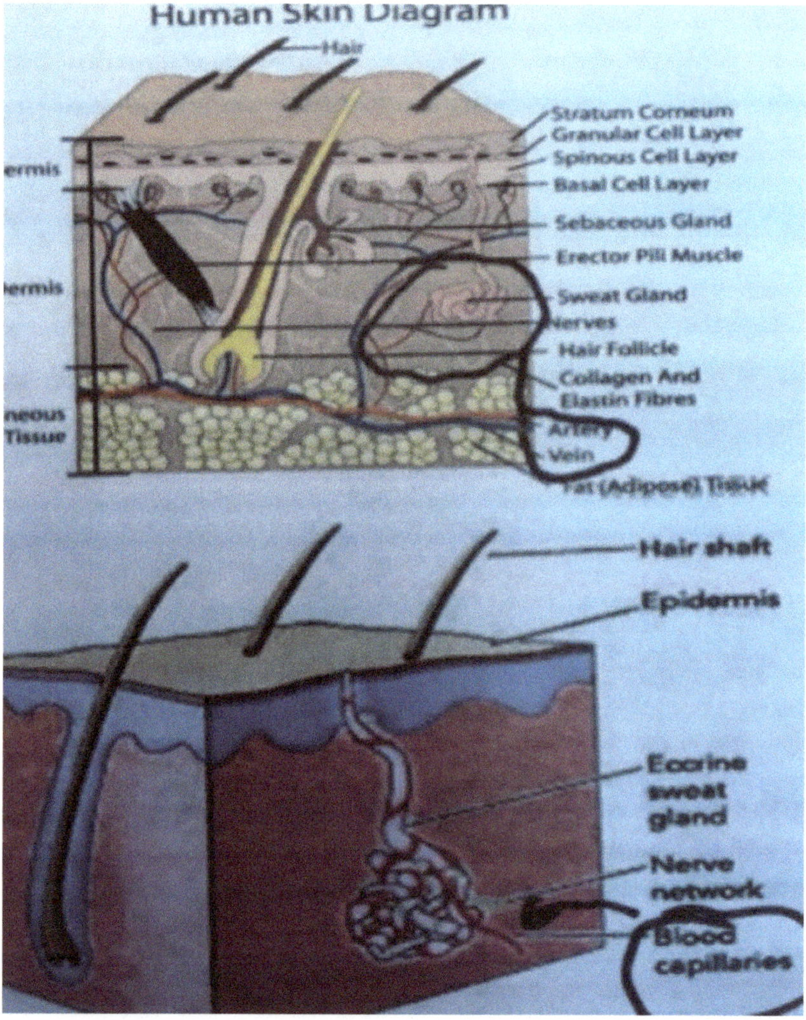

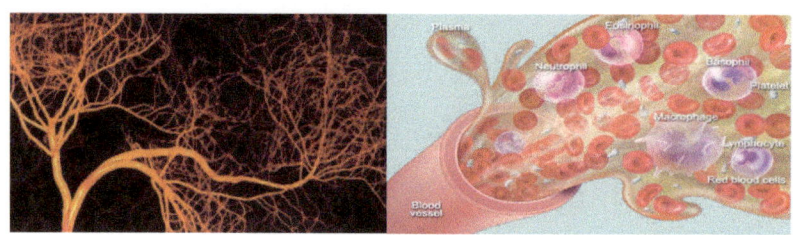

The opposite side of this large vein runs directly to the brain and has the same amount of "tree-like" distribution of blood capillaries to attach and capture cellular water transfer into the bloodstream from "Hydrostatic pressure" from raised heart rate and subsequent increased blood speed circulation to the organ and all parts of body.

THIS MARRIAGE HAS NEVER OCCURRED OR BEEN THEORIZED BY SCIENCE TO DATE! RATHER THE BLOOD CAPILLARY CONNECTIONS PRESENTED BY SCIENCE DECADES AGO (ABOVE) SHOWED THE BLOOD CAPILLARIES FEEDING THE HOSES OF THE ECCRINE GLAND AND INSTEAD FOCUSED ON THE WRONG PART, THE NERVE NETWORK IN THE DETERMINATION OF WHY AND HOW SWEATING OCCURS IN THE BODY FROM THESE GLANDS ONLY. WE KNOW THE APOCRINE GLANDS (1ST PICTURE ABOVE) HAVE ARTERY CONNECTIONS THAT RUN DIRECTLY FROM THE HEART AND ARE COMPLETELY DIFFERENT FROM THE ECCRINE GLANDS.

ALL PARTS OF THE BODY TO WHICH THE HEART PUMPS BLOOD, WHICH IRONICALLY IS THE ENTIRE BODY HAS THESE SAME EXACT CAPILLARY CONNECTIONS RUNNING FROM THEIR CELLS TO THE ECCRINE HOSE TO ALSO DEAL WITH HYDROSTATIC PRESSURE. NATURALLY IF ONE ORGAN HAS THE BLOOD CAPILLAR CONNECTIONS TO AND FROM THE ECCRINE GLANDS THEN ALL DO BECAUSE ALL HAVE TO DEAL WIT THE HYDROSTATIC PRESSURE FROM INCREASED HEART RATE AND TRANSFER THE PRESSURE OUT THE CAPILLARY CONNECTIONS.

This is why there are 640 eccrine glands per square inch of the body, along with thousands at strategic areas like the forehead, front neck and back neck to have a higher quantity of outlets for the larger and largest carriers of water in the body which are the brain, heart, and lungs!!

LITERALLY THESE BLOOD LINES ARE PART OF THE ORGAN THEMSELVES AND SEAMLESSLY DEAL WITH THE HYDROSTATIC PRESSURE IN SMOOTH FASHION WITH A SIMPLE TRANSFER FROM TREE LIKE CAPILLARIES AT THE CELL LEVEL TO RUNNING TO THE MAIN BLOOD LINE (ONE BIG LINE PICTURED ABOVE) TO THE OTHER END OF TREEL BLOOD CAPILLARY LINES TO THE ECCRINE GLANDS.

A SIMPLE YET EXTREMELY DRAMATIC FACTUAL DISCOVERY AND PART OF OUR BODY ESSENTIAL TO KEEPING ORGAN TISSUE CELLS PURE OVER TIME (DECADES)

IT'S AS EASY AS THAT! Eccrine Power & REVOLUTIONARY POWERCYCLES HAS MADE THE KEY DISCOVERY OF THE PICTURED BLOOD CAPILLARY NETWORKS MARRYING THE BRAIN AND ECCRINE GLANDS TOGETHER FOR NECESSITY OF PRESSURE RELEASE IN THE FORM OF SWEAT. THIS IS EXACTLY WHY YOU SWEAT MORE AND MORE AS YOU RAISE YOUR HEART RATE TO LEVELS LIKE 150 OR 170 BPM'S PER MINUTE (BEATS PER MINUTE). THE HIGHER YOUR HEART RATE, THE FASTER THE ENSUING BLOOD CIRCULATION TO THE ORGAN, THE MORE PRESSURE RELEASED FROM THE ATTACHED CAPILLARIES AT THE BRAIN CELL LEVEL TO THE ECCRINE GLANDS OUTLETS.

THIS BOOK HAS LITERALLY UNVEILED A KEY PART OF OUR WATER BASED BODIES TOOL FOR EASY PURIFICATION!

VERY IMPORTANT

Since it is Science that lends credibility to this entire discovery by offering FACTS that they have already booked relating to this all:

1. The body is approximately 2/3 water and the average male has up to 42 liters of water in total!

2. The largest carriers of water in the body are 1) Cells (make up the entire body) 2) Blood (travels throughout the entire body) 3. Brain 4. Heart 5. Lungs 6. Stomach 7. Liver which ARE ALL the top carriers of blood & cells (Top 2).

3. HYDRO-STATIC PRESSURE- is a term by Science that describes the process of cellular water (from cells) transfer to the bloodstream from raised heart rate and blood circulation from movement (exercise) or heat (stress to increase heart rate). Regardless of how, when the heart rate is elevated to increase faster blood circulation it results in this hydrostatic pressure to which the force of this blood extracts by movement and pressure actual water from the cells into the connecting bloodstream or blood capillaries which are the tiniest of blood vessels. This transfer was studied by Science over 4 decades ago... This is EXTREMELY relevant since it turns out to be the PERFECT marriage to my 2015 original book release and claim that sweating from running was cellular water! More importantly, this is the missing domino piece to how the blood capillaries attach microscopically to the tissues of ALL, and I mean ALL tissue cells of the body for the transfer of this cellular water upon the event of "Hydro-static pressure". Amazingly, Science never said anything about where that water is going to once in the bloodstream after the transfer, and that is only because their counterparts who creating the Eccrine glands showed the blood capillaries fusing into the house of the pore but just never theorized or connected the blood lines from the eccrine glands to those of the organs of the body.

4. There is trace amounts of detriments in this sweat also LISTED by science in TRACE amounts to include known causes to diseases such as Sodium (top ingredient), heavy metals, and REVOLUTIONARY POWERCYCLES first to announce the presence of sugars and fats in this extracted sweat.

5. The entire body already receives blood from the heart and the entire body must deal with the resulting hydrostatic pressure when occurring.

6. EACH VITAL Body part, Water percentage is as follows; Brain, 80–85%; Kidneys, 80–85%; Heart, 75–80%; Lungs, 75–80%. WOW! Let me clarify that for you all... Science already says those organs are on average over (3⁄4) three-quarters water!!!

7. Each listed organ has up to 3 pounds or greater of water in it!

8. Blood Capillaries are already Scientifically concluded to be attached to all organ tissues throughout the ENTIRE BODY! **Here is that fabulous picture again:**

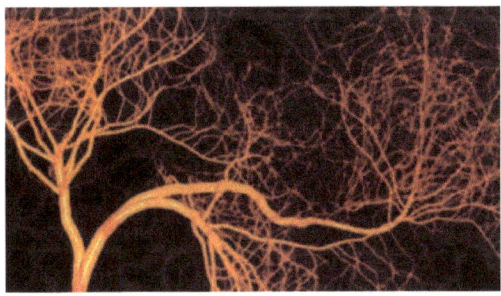

*Quite possibly the most stunning part of blood capillaries picture is how they tree down from millions of microscopic endpoints to ONE, it's worth repeating, ONE line which turns out to be COMPELLING when attached as a latest and the greatest photo which can be superimposed onto the ECCRINE pores pictured above to deliver the common senses as to how each ends TREE DOWN to one line makes the entire process work seamlessly in the body.

9. Science has already documented 640 eccrine pores per square inch of the human body!! As well as hundreds of

thousands in a limited few area of the body which strangely enough include hundreds of thousands on the human forehead (directly in front of brain), hundreds of thousands on the front of lower neck (near heart), and hundreds of thousands on the back of lower neck (near lungs). THESE ARE SCIENCE FACTS! Not things I made up... Furthermore, you can now take the time to imagine the above blood capillaries line plugging into EACH AND EVERY ONE of those eccrine pore hoses and imagine how the single line runs back to a water balloon size organ with the SAME EXACT blood capillaries going from one line to millions of attached to tissues... AND THAT, LADIES AND GENTLEMEN is how SWEATING really occurs in the human body at the eccrine glands... THROUGH pressure and force resulting from increased heart rate and circulation, resulting hydrostatic pressure and cellular water transfer to blood stream (blood capillaries) and out the pores (Eccrine).

10. Brain

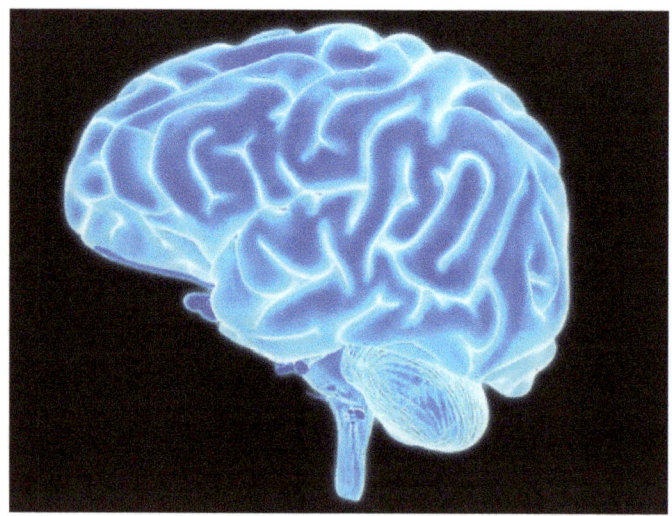

Eccrine Power

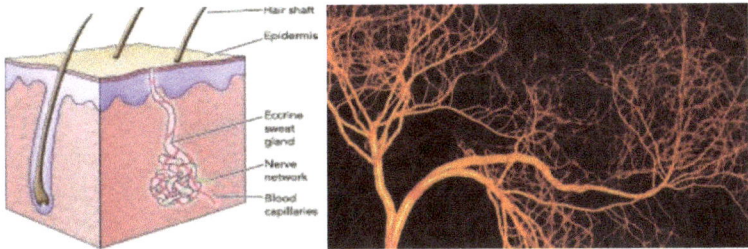

11. Liver- Another massive carrier of water, hence, cells and blood to which depend on raised heart rate and hydrostatic pressure to purify its cells base!!

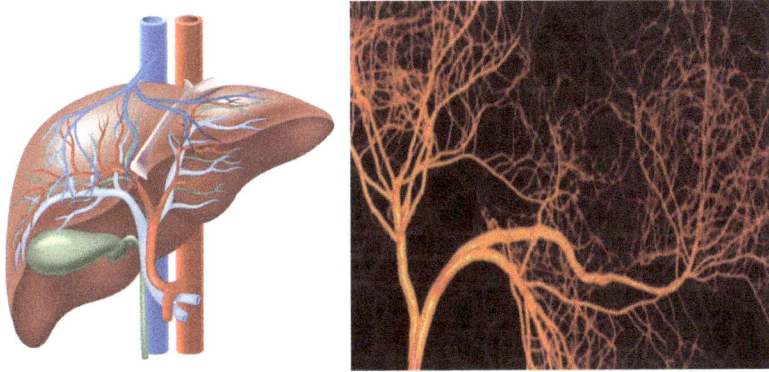

12. Breasts- Obviously, women can have more water depending on the variables... Regardless, these purify the same way as any other part of the body with cells and receive blood flow from the heart with separate blood lines coming from the heart.

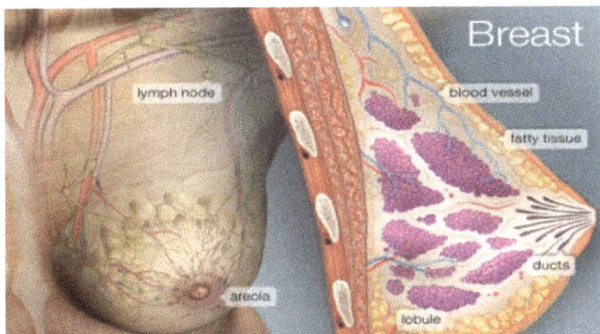

Kidneys- Another vital organ, heavily water based with tissue cells and blood; needing PURIFICATION THROUGH THESE BLOOD CAPILLARY CONNECTIONS AT THE KIDNEYS AND RUNNING DIRECTLY TO THE ECCRINE PORES!

So, the rest truly does come down to common sense and logic... If Science already says that the human heart pumps blood to the entire body and all organs and that these organs are the largest carriers of water in the body and that when the heart rate is raised due to exercise that increased blood circulation to THAT entire body again is now faster and causing real microscopic pressure dubbed "hydrostatic pressure" and which cellular water is pressed and transferred to the blood, well by golly, their own blood capillary picture fits perfectly to a TREE DOWN approach which you got it....Travels to the eccrine hose where it branches back out to release this vital process of water...

This is simple, yet powerful. Basic, yet complex. Common areas, but unique most of all by design! Or greater design by one above... oh, it's as simple as that and it's a fact since the theory of evolution is debunked in the book by the simple fact that only horses and humans have these eccrine glands powered by blood lines. As mentioned in the RP book, Apes do not have these types of sweat glands and it would be impossible for them to "SPAWN" out of nowhere due to their unique microscopic size (blood capillaries).

Don't think that's a big deal? Oh yeah it is if you can't have evolved from something since there's something basically different such as 1 million microscopic circulatory systems, connecting to millions of open hoses. WE WERE CREATED! There is no other conclusion to concur. We were given a water-based body with mobile limbs to be BUILT to RUN, BUILT to SWEAT, and BUILT to Self-Heal by recirculating the water base of our precious cells and blood to rid

Eccrine Power

them of bad detriments and toxins that AGE THEM and cause disease long term. SIMPLE as that, like a pool, the filter MUST be run to stay clean and healthy!

I DON'T FEAR! INSTEAD, I RUN AND PRAY! AND SELF-HEAL!

Use your water-base to your advantage! It's there to work on demand by pressure that you control in a simple process- MOVE YOUR LIMBS! Then Hard-press out water and bad stuff like known toxins to purify your body, mind, and soul!

This section is the conclusion of this jaw-dropping discovery of health science. The following are REAL LIFE PICTURES WHICH PERFECTLY DEMONSTRATE AND DEPICT HOW THESE BLOOD CAPILLARY LINES ARE SEAMLESSLY ATTACHED TO THE SURFACE OF AND ORGAN TO RECEIVE THE BYPRODUCT OF Hydro-static pressure at the cellular level which is the transfer of cellular water into the bloodstream. Again, this has been already studied and concluded by Science for over 4 decades!! They just never said where it goes or connect the main pipelines from the blood capillaries to another main pipeline of blood capillary trees running straight to the Eccrine pores! Simple as that...

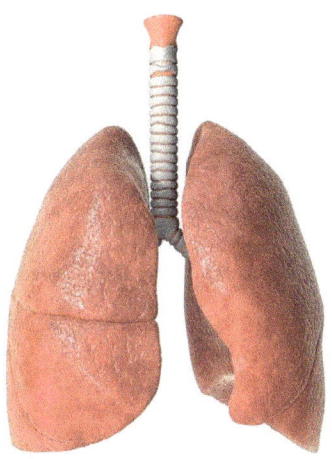

REAL Lungs pictures and actual depictions...

REAL Brain pictures and actual depictions of blood capillaries...

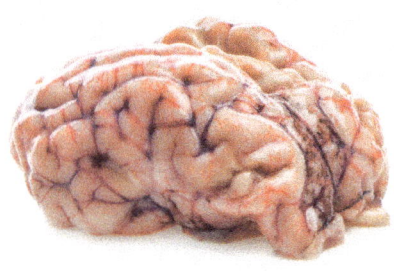 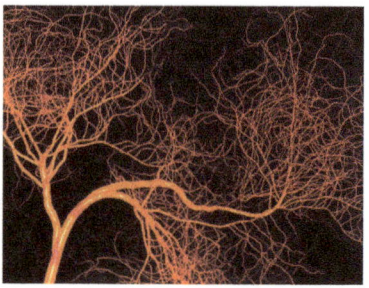

Now are you starting to see the real-life impacts of NEVER before theorizing those real blood lines as it relates to the blood capillary tree diagram below and when the below diagram is attached to the brain and any other organ, it trees to a main pipeline to and out the eccrine pores in a seamless process from GOD!

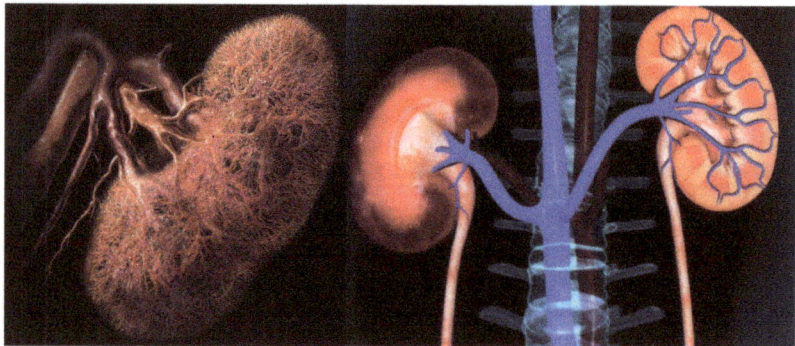

I REST MY CASE!!

My biggest concern when I was talking about blood capillaries attaching to the brain was like how are people going to believe this...?? Afterall, of all body parts, Science seems to know everything about the brain. My imagination would be visioning these on the surface. Then when I made the real-life picture discovery I almost flipped in my seat. That's all that's WAS ON the actual brain surface were blood capillaries! All there to receive cell water from increased heart rate. Nothing can be more convincing than an actual picture! Blood capillaries on the surface of the brain that tree down and run to the pores of the forehead, which are in the thousands a lot more than 640 per square inch on the rest of your body naturally to support the heart. These pores are part of the organ since the blood line runs and connects to it and goes right to the surface of the brain. This is brand new data & discovery!

I said to myself, I guess it isn't so far-fetched after all. That's exactly what the case is. This is not what I thought. I thought the surface of the brain was white and clear and nothing visible. Boy was I wrong, again how was it going to sell to the public that there's blood caps attaching to the brain seemed impossible? Yet, that's exactly the case!

Again, one of my biggest dilemmas going into these recent studies of this all, was how am I going to tell the public that these blood capillary layers are attached to the brain. It sounded so far-fetched they would think so too. And then Wala, here pops up a real-life photo of a brain that has them predominantly showing on the surface. Unbelievable!

In the health world, there's sometimes questions on whether you should raise your heart rate really high. This entire case study will prove that you absolutely must be doing so... and regularly!

"With Eccrine power, I don't get older, I get newer every day through this simple, hardworking process and through the manipulation of fast, moving water to my benefit. I'm not selling you an overnight solution, I'm selling you a lifetime solution" Giacomo Fasano. If you could run at 150 years old and look like an Olympic star just by doing a long run a bunch of times a week starting now in 2024, wouldn't you sign up? I hope so, because it's the SCIENCE that matters anyway and it dictates this to you by fact! This process is Life critical for me currently. I suffer on required days off from cardio or for me Running due to my body's addiction to the relief it provides daily for regulation of intake of diets, salts, sugars, fats, and toxins. I literally have to use my mind and try hard to battle through a day without the Eccrine power. Then the next day it's a wait till I finally get the chance to reunite with this love and feel heavenly amazing at the end! Now approaching 15+ years of this five to six day a week running habit has made me feel incredible at the very youthful age of 55! For me, it has not only stalled my Aging, but has worked to allow me to transcend looks to new heights. It can be for you or anyone as well. Just takes a little commitment and time, years to be exact.

Eccrine power and Powercycles are the soundest investment; anyone can make right now in themselves and in their future. For a simple reason that we're talking about the core health of your body and its core inner water-based components; your heart, lungs, brain, your blood and CELLS! These are all involved in this process and involved to benefit from this process. This is absolutely critical and it is not a joke. It is a simple view of the body as it was designed to be with the water base, and when you look at all the organs that are all huge carries of water they literally rely on this process and that's why they have blood connections to the pores You will benefit immediately you will benefit, almost midterm and long-term.

This incredible discovery lends much credence to the reason why our creator would make his creatures water-based. It's for a simple reason a purification of the body by pressure and movement, which it was designed for and as clearly such apparent. Time and time again, I've stated; We are built to run, built to sweat, and built to self-heal through the power of our water-based body! God chose water because when moving it's the most powerful force in nature! This is how we are made, and its recirculation has never seemed so easy then now with this amazingly simple yet very complex and intertwined process. Make no mistake about it, the entire body rests on these outlets. They are part of the organs they attach to themselves as vital connections for both purity and release! This is the right angle to look at it for success long term with our health!! END OF STORY! It's the water base, and how you should purify by high intensity movement and the use of our own powerful blood circulation amped up to cause pressure and release. It's as simple as that!

Unfortunately for those pharmaceuticals who make drugs for all of them, but very fortunately for the human race, this Jetstream-Power Cycle is the bodies, all-natural way to regulate its levels of sugar, salts, heavy metals, and fats throughout the cells and blood of the whole body. It's as simple as that! All natural to deter all those problems directly associated with their build-ups in the body, i.e., cancer, heart disease, diabetes, high blood pressure, etc... Can anybody think of them larger than life statement than that. Now you know why I'm talking about Nobel prize material here! Once again, THIS IS ECCRINE POWER! The absolute freak of nature de-Ager based on removing what causes that very word, AGING. It was never more than we already know it is!! Toxins and Detriments!

Well this has always staked claim to being able to defeat them all for the very simple reason that this power is fed by none other than direct blood feeds from blood capillary lines that Tree down in formation. Perfect! What an act of God, to attach these microscopic blood lines to and from the organ's tissues to the outlets of the world... the Eccrine tunnel duct pore/hose...!

Your water-based body is not a garbage can it must be purified through this process or its water-base will thicken and resonate with all the harmful detriments that caused it harm over time.

Simple perspective, all the angle of the amount of water in our body as a whole has to be the determining factor why we were given these outlets to purify Eccrine powers, huge discovery comes down to the water-based bodies, a critical and essential purification process that's been overlooked by the simple fact of NOT theorizing the bloodlines that feed into the pores! traditional Science missed these and focused on the surroundings nerve network, and said, sweating is a way for the body to cool down when hot, but the Eccrine pores have nothing to with this actuality! It's literally all and ONLY heart rate and increased blood flow to and from the organs to the blood capillaries seamlessly connecting to the entire hose package of the Eccrine pore. Simply Incredible. Water on the surface naturally cools the body from the open air and it's only flowing out due to this increase in bpm's (beats per minute)

Let me tell you the biggest news of them all. I am the first person to EVER in the world to identify and theorize these blood capillaries attaching to the surface of the brain! Yes. the first! They are ONLY there to fill with cellular water resulting from HYDROSTATIC PRESSURE (Science defined study for cellular water transfer from cells to blood from increased heart rate and blood flow to organ). Yes, water and mostly only water, little blood and when this process occurs the water simply travels downward the blood capillary trees to a single line where it runs directly to another blood capillary tree which connects to thousands of eccrine pore sac bases to be emitted as sweat-water! Guess what, those blood lines will never fill with water if the heart rate never increases from movement to cause the necessary pressure to squeeze out the water from the cells to the bloodstream (Hydrostatic-pressure). Please note and do not forget,

Science has already coined the phrase hydrostatic pressure and completed a study over 4 decades ago on how it results from exercise and increased heart rate and blood flow. Why oh why, or how and how did they not connect the dots or fill in the blanks on these bloodlines is beyond me?? Likely, the reason is the MISSED theory of the blood capillaries at the eccrine pore base as mentioned prior. Another senseless and vital miss that opened the door for me and my backing into this all through a self-obtained obsession with running and sweat 15 years ago!

I'm waiting for all the world's renowned Scientists to pick up the phone and try to challenge ANY PART of my REAL-LIFE findings... Try to tell me what you think those blood capillaries are doing on the surface of the brain and ALL other organs internally and why you can't figure out they run directly to an Eccrine pore base to be extracted... Tell me, we are all waiting! It's ok, we'll do it without you! The general public will prevail as we excel into the future of health and long-term lives over 150 years and beyond! The Skipping Powercycle

Glossary of Scientific Terms and Facts

1. Hydrostatic Pressure

"The primary purpose of the cardiovascular system is to circulate gases, nutrients, wastes, and other substances to and from the cells of the body. Small molecules, such as gases, lipids, and lipid-soluble molecules, can diffuse directly through the membranes of the endothelial cells of the capillary wall. Glucose, amino acids, and ions—including sodium, potassium, calcium, and chloride—use transporters to move through specific channels in the membrane by facilitated diffusion. Glucose, ions, and larger molecules may also leave the blood through intercellular clefts. Larger molecules can pass through the pores of fenestrated capillaries, and even large plasma proteins can pass through the great gaps in the sinusoids. Some large proteins in blood plasma can move into and out of the endothelial cells packaged within vesicles by endocytosis and exocytosis. Water moves by osmosis."

2. Bulk Flow

"The mass movement of fluids into and out of capillary beds requires a transport mechanism far more efficient than mere diffusion. This movement, often referred to as bulk flow, involves two pressure-driven mechanisms: Volumes of fluid move from an area of higher pressure in a capillary bed to an area of lower pressure in the tissues via filtration. In contrast, the movement of fluid from an area of higher pressure in the tissues into an area of lower pressure in the capillaries is reabsorption. Two types of pressure interact to drive each of these movements: hydrostatic pressure and osmotic pressure.

Hydrostatic Pressure

The primary force driving fluid transport between the capillaries and tissues is hydrostatic pressure, which can be defined as the pressure of any fluid enclosed in a space. Blood hydrostatic pressure is the force exerted by the blood confined within blood vessels or heart chambers. Even more specifically, the pressure exerted by blood against the wall of a capillary is called capillary hydrostatic pressure (CHP), and is the same as capillary blood pressure. CHP is the force that drives fluid out of capillaries and into the tissues. As fluid exits a capillary and moves into tissues, the hydrostatic pressure in the interstitial fluid correspondingly rises. This opposing hydrostatic pressure is called the interstitial fluid hydrostatic pressure (IFHP). Generally, the CHP originating from the arterial pathways is considerably higher than the IFHP, because lymphatic vessels are continually absorbing excess fluid from the tissues. Thus, fluid generally moves out of the capillary and into the interstitial fluid. This process is called filtration.

3. Osmotic Pressure

The net pressure that drives reabsorption—the movement of fluid from the interstitial fluid back into the capillaries—is called osmotic pressure (sometimes referred to as oncotic pressure). Whereas hydrostatic pressure forces fluid out of the capillary, osmotic pressure draws fluid back in. Osmotic pressure is determined by osmotic concentration gradients, that is, the difference in the solute-to-water concentrations in the blood and tissue fluid. A region higher in solute concentration (and lower in water concentration) draws water across a semipermeable membrane from a region higher in water concentration (and lower in solute concentration).

As we discuss osmotic pressure in blood and tissue fluid, it is important to recognize that the formed elements of blood do not contribute to osmotic concentration gradients. Rather, it is the plasma proteins that play the key role. Solutes also move across the capillary wall according to their concentration gradient, but overall, the concentrations should be similar and not have a significant impact on osmosis. Because of their large size and chemical structure, plasma proteins are not truly solutes, that is, they do not dissolve but are dispersed or suspended in their fluid medium, forming a colloid rather than a solution.

The pressure created by the concentration of colloidal proteins in the blood is called the **blood colloidal osmotic pressure (BCOP)**. Its effect on capillary exchange accounts for the reabsorption of water. The plasma proteins suspended in blood cannot move across the semipermeable capillary cell membrane, and so they remain in the plasma. As a result, blood has a higher colloidal concentration and lower water concentration than tissue fluid. It therefore attracts water. We can also say that the BCOP is higher than the **interstitial fluid colloidal osmotic pressure (IFCOP)**, which is always very low because interstitial fluid contains few proteins. Thus, water is drawn from the tissue fluid back into the capillary, carrying dissolved molecules with it. This difference in colloidal osmotic pressure accounts for reabsorption.

4. Interaction of Hydrostatic and Osmotic Pressures

The normal unit used to express pressures within the cardiovascular system is millimeters of mercury (mm Hg). When blood leaving an arteriole first enters a capillary bed, the CHP is quite high—about 35 mm Hg. Gradually, this initial CHP declines as the blood moves through the capillary so that by the time the blood has reached the venous end, the CHP has dropped to approximately 18 mm Hg. In comparison, the plasma proteins remain suspended in the blood, so the BCOP remains fairly constant at about 25 mm Hg throughout the length of the capillary and considerably below the osmotic pressure in the interstitial fluid.

The **net filtration pressure (NFP)** represents the interaction of the hydrostatic and osmotic pressures, driving fluid out of the capillary. It is equal to the difference between the CHP and the BCOP. Since filtration is, by definition, the movement of fluid out of the capillary, when reabsorption is occurring, the NFP is a negative number.

NFP changes at different points in a capillary bed. Close to the arterial end of the capillary, it is approximately 10 mm Hg, because the CHP of 35 mm Hg minus the BCOP of 25 mm Hg equals 10 mm Hg. Recall that the hydrostatic and osmotic pressures of the interstitial fluid are essentially negligible. Thus, the NFP of 10 mm Hg drives a net movement of fluid out of the capillary at the arterial end. At approximately the middle of the capillary, the CHP is about the same as the BCOP of 25 mm Hg, so the NFP drops to zero. At this point, there is no net change of volume: Fluid moves out of the capillary at capillary, the CHP has dwindled to about 18 mm Hg due to loss of fluid. Because the BCOP remains steady at 25 mm Hg, water is drawn into the capillary, that is, reabsorption occurs. Another way of expressing this is to say that at the venous end of the capillary, there is an NFP of −7 mm Hg.

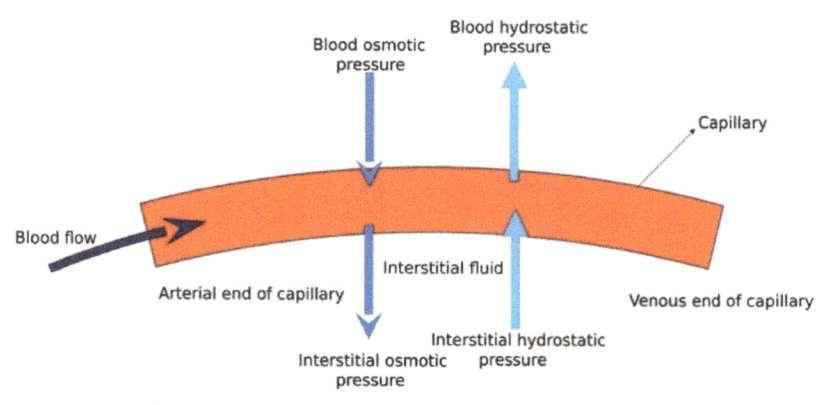

Figure 1. Net filtration occurs near the arterial end of the capillary since capillary hydrostatic pressure (CHP) is greater than blood colloidal osmotic pressure (BCOP). There is no net movement of fluid near the midpoint since CHP = BCOP. Net reabsorption occurs near the venous end since BCOP is greater than CHP.

The Role of Lymphatic Capillaries

Since overall CHP is higher than BCOP, it is inevitable that more net fluid will exit the capillary through filtration at the arterial end than enters through reabsorption at the venous end. Considering all capillaries over the course of a day, this can be quite a substantial amount of fluid: Approximately 24 liters per day are filtered, whereas 20.4 liters are reabsorbed. This excess fluid is picked up by capillaries of the lymphatic system. These extremely thin-walled vessels have copious numbers of valves that ensure unidirectional flow through ever-larger lymphatic vessels that eventually drain into the subclavian veins in the neck. An important function of the lymphatic system is to return the fluid (lymph) to the blood. Lymph may be thought of as recycled blood plasma.

(Seek additional content for more detail on the lymphatic system.)"

Reference: https://courses.lumenlearning.com/suny-ap2/chapter/capillary-exchange/

Small molecules can cross into and out of capillaries via simple or facilitated diffusion. Some large molecules can cross in vesicles or through clefts, fenestrations, or gaps between cells I capillary walls. However, the bulk flow of capillary and tissue fluid occurs via filtration and reabsorption. Filtration, the movement of fluid out of the capillaries, is driven by the CHP. Reabsorption, the influx of tissue fluid into the capillaries, is driven by the BCOP. Filtration predominates in the arterial end of the capillary; in the middle section, the opposing pressures are virtually identical so there is no net exchange, whereas reabsorption predominates at the venule end of the capillary. The hydrostatic and colloid osmotic pressures in the interstitial fluid are negligible in healthy circumstances.

Glossary

Blood Colloidal Osmotic Pressure (BCOP): pressure exerted by colloids suspended in blood within a vessel; a primary determinant is the presence of plasma proteins.

Blood Hydrostatic Pressure: force blood exerts against the walls of a blood vessel or heart chamber.

Capillary Hydrostatic Pressure (CHP): force blood exerts against a capillary.

Filtration: in the cardiovascular system, the movement of material from a capillary into the interstitial fluid, moving from an area of higher pressure to lower pressure

Interstitial Fluid Colloidal Osmotic Pressure (IFCOP): pressure exerted by the colloids within the interstitial fluid interstitial fluid hydrostatic pressure (IFHP): force exerted by the fluid in the tissue spaces.

Net Filtration Pressure (NFP): force driving fluid out of the capillary and into the tissue spaces; equal to the difference of the capillary hydrostatic pressure and the blood colloidal osmotic pressure

Reabsorption: in the cardiovascular system, the movement of material from the interstitial fluid into the capillaries

https://courses.lumenlearning.com/suny-ap2/chapter/capillary-exchange/

Ok, so why are these "ultra science quotes" pasted for your reading? Simple, so you all understand the complexity of "Known Science" regarding our very own circulatory system and specifically the blood capillaries and hydro-static pressure and Capillary hydro-static pressure which is what the tissues connect to as mentioned in prior chapters as well as at the ECCRINE PORE levels which are the pressure-built exit points... You needed Science to convince you of my wonderful findings. Well there it is in plain English!

If I tell you that the pressure that Science already studied is the driving force behind my Eccrine Powercycle and the bodies built in way to "force out bad", pragmatically and efficiently and that this angle has never been considered then your COMMON SENSE of the visuals and facts are all you need to believe in this God given system. Nevermind the fact that I've been personally studying it and performing it for over 15 years straight with REAL RESULTS which means YOU should listen to me and JUST START DOING THE PRESSURE BUILT **ECCRINE POWERCYCLE!!** Diet, Yoga, Stretching, and Strength training

Sample Diets:

Breakfast:

1. Eggs are the most important protein on the planet! Yes, the planet. I don't miss a day. They are literally the only Source of protein to magically contain ALL of the Amino acids of proteins. All! Nothing else can stake the claim. When I switched to pasteurized, open range eggs, I literally saw a noticeable difference in my hair and nails growth. This was because eggs already have the entire spectrum of said Amino acids, but when I ate the fully organic eggs the nature of this formula exploded on me and I realized the real importance of NON-GMO eggs... They are stripped of nutrition. Anyhow, EAT YOUR EGGS DAILY! I'm a sunny-side up fan... lol

2. Organic Coffee and fruit- A big difference maker on your body's ability to recover and

recharge.

3. Organic Oatmeal

4. Toast and butter

5. Fruits

Lunch

1. Chicken- great source of protein that does not weigh the body down prior to workouts.

2. Beef and beef fat- both are critically vital to the exercising water-based

body!

3. Pasta, Rice, and potatoes- Big 3 Carbohydrates for energy!

4. Organic fruit- Apples #1, all types must be organic! bananas and oranges

Dinner

Top Proteins for dinner include: Crabmeat (fresh catch), Prime rib, Chicken, Turkey, and Tuna.

Top Carbohydrates for dinner include: White rice, pasta, Idaho potatoes, and Garlic (fresh cut-organic

only) bread

Top Vegetables for dinner- Mixed medley Organic broccoli, cauliflower, and carrots. Green beans, Red

peppers, mushrooms, and tomatoes.

Snacks (In between meals)

Organic peanut butter and jelly

Cut Apples (Organic only!)

Health bars- Perfect bars (9), Cliff bars (8), Kind bars (8)

Bananas (Organic only)

New Prayer for Eccrine Power!

Prayer of Bodily Love

by Giacomo Fasano

Oh, heavenly father. Oh great God, you have given me this beautiful body, one that's mobile and agile, one that can climb and run, skip and walk, grab and feel. Oh Lord, you are the greatest. You are the smartest with all power to you. We appreciate all your gifts, your intelligence, and your ability to keep us safe, we adore the body you designed for us. We manifest it! We want to work it out to your expectations. We want to sit in line with you and feel holy with God spirit so we may serve you now and in the afterlife! But now let us have a great life, long & healthy filled and completely committed to your great works. Let us purify our bodies, minds and spirits through the use of our entire being that you gave us. Thank you, Lord.

Prayer for Life's Appreciation (written for Revolutionary Powercycles in 2015)

Oh, good Lord, our heavenly Father, I pray to you today with all my hope and faith. Vested in your mysterious, but great ways, I don't question or judge, rather always only turn to you with all my heart and passion for a nudge towards the right direction.

When I experience doubt or even temptation from the evil one, I galvanize my heart and soul toward you, so together we may rise above once again, and demonstrate your righteous ways.

Fore even if at times it may seem a bit dim; your path is always lit and visible to my eyes. Whereas, I know the other direction is filled with deception and attractive colors that will only fade and dissipate in dismay.

Where we will not always understand, let us never fail to feel your glory and everlasting presence in our life. Pray down on us, as we pray for you, and always strengthen our fortitude for your love and image. Thank you, God for all the joy and gifts of Life you provide every day.

Amen

Religion:

Time and faith have a strange way of healing many wounds. Time and prayer also have a strange way of healing ones in our soul.

I get it! Running is boring. It's so hard to start not even the first time but every time even after 20 years starting to run something you just don't wanna do but you just have to do it go through the motions once you start running you literally don't wanna stop just remember that.

The Skipping Powercycle

The most unknown exercise on the planet is no doubt skipping. What is skipping officially? Powercycles define skipping as the combination of jumping and running at the same time for art for the body that if over an hour, creates a great sweat and work out.

So I introduced skipping powercycle one hour plus of jumping, running together or skipping, another bodily function designed by God and unique of our limbs and body

Running for Dummies (like Windows for dummies)

I have learned one BIG tip to succeed in both starting and maintaining a good running habit long term. The preface to it is simple as well... Listen to your body! Then my 20+ years have taught me that BOTH, yes both Novices and Pros follow this very straight forward strategy to run for an hour and plus... IT IS:

RUN TILL YOUR TIRED, WALK TILL YOUR NOT! Then Run again till you're tired (Also listening to your body) and Walk till you're not (tired). The reason this rule applies to both beginners, advanced, and experts is simple; because the only variable is how long you need to walk to recoup. As time goes on and you go out more for long runs and walks, the gap CONSTANTLY shrinks but the process ALWAYS stays the same,

Run till you're tired, then stop and Walk till you're not, Run again till tired and stop and walk till you're not. Repeat. To this day It's what I do but now just for minutes over an hour plus due to my earned and acquired conditioning. So Betty the beginner may listen to me and hit the track for an hour and the first few times end up only running for 10 minutes of the 60 and walking the majority of 50 minutes... However, she will HAVE in fact run and walked an hour and compiled a range of 2 1.2 to 4 miles in doing so... Then we see Betty in her third month averaging 30 minutes of run time and 3o minutes of walk time in that same hour! Yes, then in just 6-9 months averaging 40 minutes of run time and 20 min of walk to show the natural progression. All the while, doing the same Powercycle technique of Run till you're tired, then walk till you're NOT! Repeat!

This is literally just an easy formula to allow anybody that can walk to run in for hours and days. No problem. The formula is actually rather simple and works exactly the same for a beginner and an expert. Believe it or not yes it works the same for both, because no matter if you're an expert or novice or medium experience runner, it all comes down to this, You run to your tired and you walk until you're not you run again till you're tired you stop and walk till you're not repeat and that's how you run for hours to allow yourself to stop and recoup naturally as you run that gap decreases on walk time, and it becomes more running time But the same principle always applies to your tired walk till you're not Can I get any

simpler than that? Not really you just need to walk that run stop on your tired walk till you recoup repeat

In conclusion, you must understand one simple thing. Your body's net water composition of 66% requires you to use and benefit from its sole purification process! You simply must find a way to either run and walk for an hour routinely or other high-intensity cardiovascular activities sustained for over 60 minutes as well to complete this vital process. You will succeed to feel healthier and with more overall vitality!! You will feel and look better, sleep better, think better, have unlimited energy and most importantly FEEL a significant uptick in your Faith and Spirituality. This is a must do, not nice to do. From its most basic perspective, the Powercycle of the Eccrine pores is all common sense! The conclusion is that the heart already pumps blood to all these resting organs at a normal heart rate or bpm's of 65-85! This study does nothing more than analyze what happens when the bpm's are raised from movement (or heat) to elevated levels of over 110-170 bpm and the resulting pressure that the tissue cells of ALL BODILY ORGANS must deal with—henceforth the blood infused outlets (Eccrine pores) that are 640 per square inch of the body and hundreds of thousands on the strategic areas of the forehead, frontal and back of neck. AMAZING! Or more like Amazing Grace!!

Listening to me will create a perfect balance. I've dubbed the **3 H's...Happy, Holy, & Healthy** that will be promised and delivered for all that can sustain the Powercycle and use their own bodies "Eccrine Power" for just over 1 year!! That's all it takes to feel the magic of a "flushed out internal body with optimal water recirculation"

"Run & Sun"

Not all running; in fact, most doesn't occur outside in our world; I should say it occurs a lot on the treadmill. However, when the weather cooperates, there is certainly nothing more holistic than the actual run in the sun Eccrine Power!

As I advanced over the years, I learned the better shape I was getting in, the more I could enjoy running in the sun. taking your shirt off if you're a guy & adding to your tan...

Talk about a FULL body charge! absorbing Vitamin D, sweating out Toxins, burning calories, releasing endorphins, feeling Adrenalin and aligning your spirit with meditation and simple prayer. Incredibly leashes the fountain of the eccrine power flow!

And remember the simple Success formula for beginners or experts is always JUST walk at first, then run till you're tired, walk again till you're not tired, run again till you are tired, stop walk again till you're not tired, and then run again and repeat. This is how you attain your time goal of 60+ minutes, by incorporating the walk and run combination and not quitting the time for it is at the end when all the rewards are reaped and only then for most cases OK 45 min or greater is going to give you a good taste of this power but when you go past the hour, it is the ultimate Enrichment! Look at it this way. If your family was in danger and they were 5 miles away and you had to get there and on foot to help and it was 90 minutes away. This is what you would do. You would run till you can't and you would stop And walk and run, but you wouldn't quit the time because you wanna get there to save your family. The urgency needs to be there that's how you attain long runs. Walk, Run, walk & run.

Eccrine Power

Top 10 Run in the Sun songs.

1. "Calling" – Alesso
2. "Sending all my love" – Linear
3. "Love comes waking in" – Van Halem
4. "Why can't this be love" – Van Halen

    Best lyrics "You got to Run to win"

    (Van Halen double)

5. "All my Love" – Led Zeppelin
6. "How can we be wrong" – Trinere

Bonus Best Classic Run songs

45

7. "Grease" (from Grease) – Frankie Vallie
8. "You're the one that I want" (Grease) – Olivia Newton-John

   Best lyrics "Meditate in my direction…"

   (Grease double)
9. "Can't take my eyes off of you" – Ms. Lauren Hill
10. "Everything is Everything" – Ms. Lauren Hill

Quick Note-Always change the person or character in a song to God to make the lyrics work towards him during prayer and meditation!

Eccrine "Star" Power

Amazingly, superstar, running athletes in soccer and basketball like LeBron James, Caitlyn Clark, Cristiano Ronaldo, and Lionel Messi all use that power without even knowing it as an underlying benefit of being paid to RUN!

They may think their salaries and contracts are worth a lot, but in fact, it's the Eccrine Power that they are using, secretly and unbeknownst to them that's worth more than their PAY, for health is wealth and look at them. This could be you just use Eccrine power for years and it will happen!!

Not much more To say about this topic. It is what it is and They are endorsing me and EPower Without even trying to do it because it's underlying.

and they are professional athletes that you know who run and sweat and use The magic of Eccrine power!

Let me spend a few minutes on blood speed because its this that I believe is now the number one tool human beings have to live close to immortal. I'll say it again our blood speed is the key longevity.

Why? It's simple...

First of all, it's traveling fast!

Number two- Its traveling very fast!

Number three= Its traveling very very fast just in idle position like sitting on a couch!

Number 4- human blood is about 90% water, it's mainly made up of water, which again is moving very fast and again moving water is the most powerful substance in the world. Let's recap. Moving water like a tsunami is one of the worlds known strongest powers. Human blood is made up of almost water and is also moving fast. Therefore, even at idle heart rate, its comparable to a tsunami in its scientific aspects of moving speed of water and its ability to take anything physical matter with it. So, it is blood speed is like a human tsunami. THEREFORE, its face value holds it high as a hidden tool due to raw science!!

Number 5- This blood-water based human tsunami called our blood circulation touches EVERY single part of our insides including all our all our water-based cells!

Number 6- This blood speed is robust and powerful traveling throughout the body. Just like natures ocean tsunami; when high winds are added it can be even stronger and forceful like the human blood speed when a person runs his or her heartrate up from thrust and movement from like running.

Number 7- This controllable aspect of Power wash or when blood circulation is elevated is literally comparable to a firehose's thrust and power as the blood thrusts against human water-based cells violently from pressure created from over 130 bpm's (beats per minute) of elevated heart rate. Now the blood speed that has been turbo-boosted manually by YOU is your most Magical tool of internal cleaning.

So, blood speed is moving very fast and blood speed is blood water so its water moving very fast. Water moving very fast like a natural tsunami is amongst the most powerful forces of nature.

Conclusion: Since blood speed is majority water and moving at extremely fast speeds, it is the human version of "Moving water" power and then equally must be classified as a #1 tool for its simple basis of comparison. No, its not a stretch to compare an ocean tsunami with a human blood tsunami since the internal combustion is in fact identical! What does Science world have to say about this aspect of our amazing study??

It touches our insides at a very hard rate when we elevate that speed by running.

This is why blood speed to accelerate blood to control our health and well-being.

Anybody who argues with this doesn't have common sense.

For you cannot argue with power of moving water and the fact that blood is 90% water. It's already moving at high speeds. In my first book, Revolutionary Powercycles, I called our blood circulation "Jetstream like" for its unbelievable raw speed and force. Jetstream force is turbo speed and precision as is our blood circulation. Does anyone really want to keep this tool idle a majority of the time? Doesn't make sense. It only makes senses to use the power and force of throttling it up just like revving a sports car engine to feel the power and benefits of FORCE!

Finally, I will conclude that this violent process of kickstarting out already hyper fast blood speed to higher levels to create hydrostatic pressure all results in an odd turnout, which is the calm release of water at the surface because it goes through a press at the pore hose... Do you understand all that pressure and violence releases culminates itself in a very silently and calm nature on the skins surface. Yet, it was produced from such violence internally. This to me is the most amazing part of the Eccrine power cycle is the violence internally, and the calmness on the outside surface form of sweat. This violent to calm effect of the water smoothly coming out the surface proves that this process is all pressure built in the higher, the pressure, the more the oozing out the pores.

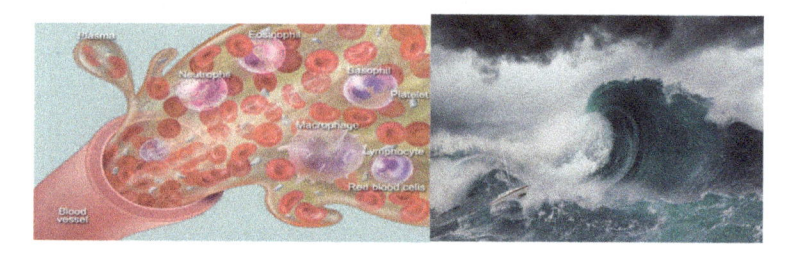

Above pictured are TWO separate all powerful Tsunamis... One of force of the natures ocean, and one the force of human blood stream. Both equal in force. One controllable only by Mother nature, and the other, CONTROLLABLE but ignored by human beings.

Rank	Substance Name	Total Points	CAS RN
1	ARSENIC	1675	7440-38-2
2	LEAD	1531	7439-92-1
3	MERCURY	1455	7439-97-6
4	VINYL CHLORIDE	1355	75-01-4
5	POLYCHLORINATED BIPHENYLS	1342	1336-36-3
6	BENZENE	1328	71-43-2
7	CADMIUM	1317	7440-43-9
8	BENZO(A)PYRENE	1306	50-32-8
9	POLYCYCLIC AROMATIC HYDROCARBONS	1277	130498-29-2
10	BENZO(B)FLUORANTHENE	1255	205-99-2
11	CHLOROFORM	1202	67-66-3
12	AROCLOR 1260	1191	11096-82-5
13	DDT, P,P'-	1182	50-29-3
14	AROCLOR 1254	1172	11097-69-1

15	DIBENZO(A,H)ANTHRACENE	1163	53-70-3
16	TRICHLOROETHYLENE	1153	79-01-6
17	CHROMIUM, HEXAVALENT	1151	18540-29-9
18	DIELDRIN	1142	60-57-1
19	PHOSPHORUS, WHITE	1141	7723-14-0
20	AROCLOR 1242	1126	53469-21-9
21	DDE, P,P'-	1125	72-55-9
22	CHLORDANE	1125	57-74-9
23	COAL TAR CREOSOTE	1124	8001-58-9
24	HEXACHLOROBUTADIENE	1122	87-68-3
25	ALDRIN	1115	309-00-2
26	DDD, P,P'-	1113	72-54-8
27	AROCLOR 1248	1107	12672-29-6
28	HEPTACHLOR	1102	76-44-8
29	AROCLOR	1102	12767-79-2
30	BENZIDINE	1093	92-87-5

31	ACROLEIN	1090	107-02-8
32	TOXAPHENE	1089	8001-35-2
33	TETRACHLOROETHYLENE	1081	127-18-4
34	HEXACHLOROCYCLOHEXANE, GAMMA-	1074	58-89-9
35	CYANIDE	1068	57-12-5
36	HEXACHLOROCYCLOHEXANE, BETA-	1054	319-85-7
37	DISULFOTON	1049	298-04-4
38	BENZO(A)ANTHRACENE	1048	56-55-3
39	1,2-DIBROMOETHANE	1044	106-93-4
40	DIAZINON	1038	333-41-5
41	HEXACHLOROCYCLOHEXANE, DELTA-	1035	319-86-8
42	ENDRIN	1035	72-20-8
43	BERYLLIUM	1033	7440-41-7
44	AROCLOR 1221	1029	11104-28-2
45	1,2-DIBROMO-3-CHLOROPROPANE	1027	96-12-8

46	ENDOSULFAN	1022	115-29-7
47	ENDOSULFAN, ALPHA	1019	959-98-8
48	HEPTACHLOR EPOXIDE	1019	1024-57-3
49	CIS-CHLORDANE	1017	5103-71-9
50	CARBON TETRACHLORIDE	1015	56-23-5
51	COBALT	1015	7440-48-4
52	AROCLOR 1016	1013	12674-11-2
53	DDT, O,P'-	1010	789-02-6
54	PENTACHLOROPHENOL	1007	87-86-5
55	ENDOSULFAN SULFATE	1005	1031-07-8
56	METHOXYCHLOR	1004	72-43-5
57	NICKEL	994	7440-02-0
58	ENDRIN KETONE	994	53494-70-5
59	DI-N-BUTYL PHTHALATE	993	84-74-2
60	DIBROMOCHLOROPROPANE	985	67708-83-2

61	BENZO(K)FLUORANTHENE	975	207-08-9
62	TRANS-CHLORDANE	969	5103-74-2
63	ENDOSULFAN, BETA	969	33213-65-9
64	CHLORPYRIFOS	965	2921-88-2
65	XYLENES, TOTAL	962	1330-20-7
66	CHROMIUM(VI) TRIOXIDE	962	1333-82-0
67	AROCLOR 1232	959	11141-16-5
68	ENDRIN ALDEHYDE	959	7421-93-4
69	METHANE	952	74-82-8
70	2,3,7,8-TETRACHLORODIBENZO-P-DIOXIN	943	1746-01-6
71	3,3'-DICHLOROBENZIDINE	942	91-94-1
72	2-HEXANONE	940	591-78-6
73	BENZOFLUORANTHENE	937	56832-73-6
74	ZINC	916	7440-66-6

75	TOLUENE	912	108-88-3
76	PENTACHLOROBENZENE	907	608-93-5
77	DI(2-ETHYLHEXYL)PHTHALATE	903	117-81-7
78	CHROMIUM	892	7440-47-3
79	2,4,5-TRICHLOROPHENOL	892	95-95-4
80	AROCLOR 1240	890	71328-89-7
81	2,4,6-TRINITROTOLUENE	879	118-96-7
82	NAPHTHALENE	878	91-20-3
83	1,1-DICHLOROETHENE	873	75-35-4
84	BROMODICHLOROETHANE	868	683-53-4
85	DDD, O,P'-	868	53-19-0
86	BIS(2-CHLOROETHYL)ETHER	868	111-44-4
87	2,4,6-TRICHLOROPHENOL	867	88-06-2
88	HYDRAZINE	863	302-01-2
89	2,4-DINITROPHENOL	860	51-28-5
90	4,4'-METHYLENEBIS(2-CHLOROANILINE)	859	101-14-4
91	METHYLENE CHLORIDE	858	75-09-2

92	1,2-DICHLOROETHANE	854	107-06-2
93	N-NITROSODIMETHYLAMINE	849	62-75-9
94	THIOCYANATE	848	302-04-5
95	HEXACHLOROBENZENE	845	118-74-1
96	ASBESTOS	840	1332-21-4
97	RADIUM-226	834	13982-63-3
98	RDX (Cyclonite)	833	121-82-4
99	URANIUM	833	7440-61-1
100	2,4-DINITROTOLUENE	832	121-14-2
101	ETHION	832	563-12-2
102	4,6-DINITRO-O-CRESOL	829	534-52-1
103	RADIUM	827	7440-14-4
104	THORIUM	823	7440-29-1
105	DIMETHYLARSINIC ACID	823	75-60-5
106	CHLORINE	822	7782-50-5
107	1,3,5-TRINITROBENZENE	820	99-35-4

108	HEXACHLOROCYCLOPENTADIENE	820	77-47-4
109	RADON	818	10043-92-2
110	HEXACHLOROCYCLOHEXANE, ALPHA-	816	319-84-6
111	RADIUM-228	815	15262-20-1
112	THORIUM-230	814	14269-63-7
113	URANIUM-235	812	15117-96-1
114	THORIUM-228	810	14274-82-9
115	RADON-222	810	14859-67-7
116	URANIUM-234	809	13966-29-5
117	N-NITROSODI-N-PROPYLAMINE	809	621-64-7
118	METHYLMERCURY	808	22967-92-6
119	COAL TARS	808	8007-45-2
120	COPPER	807	7440-50-8

121	CHRYSOTILE ASBESTOS	806	12001-29-5
122	PLUTONIUM-239	806	15117-48-3
123	POLONIUM-210	805	13981-52-7
124	PLUTONIUM-238	805	13981-16-3
125	LEAD-210	805	14255-04-0
126	1,1,1-TRICHLOROETHANE	804	71-55-6
127	AMOSITE ASBESTOS	804	12172-73-5
127	PLUTONIUM	804	7440-07-5
127	STRONTIUM-90	804	10098-97-2
130	RADON-220	804	22481-48-7
131	BARIUM	804	7440-39-3
132	AMERICIUM-241	804	86954-36-1
133	HYDROGEN CYANIDE	803	74-90-8
134	AZINPHOS-METHYL	802	86-50-0

134	CHLORDANE, TECHNICAL	802	12789-03-6
136	CHLORDECONE	802	143-50-0
137	MERCURIC CHLORIDE	802	7487-94-7
137	NEPTUNIUM-237	802	13994-20-2
139	ACTINIUM-227	801	14952-40-0
139	PLUTONIUM-240	801	14119-33-6
141	CHLOROBENZENE	801	108-90-7
142	S,S,S-TRIBUTYL PHOSPHOROTRITHIOATE	799	78-48-8
143	MANGANESE	799	7439-96-5
144	ETHYLBENZENE	798	100-41-4
145	FLUORANTHENE	797	206-44-0
146	CHRYSENE	794	218-01-9
147	PERFLUOROOCTANE SULFONIC ACID	790	1763-23-1
148	1,2,3-TRICHLOROBENZENE	786	87-61-6

149	POLYBROMINATED BIPHENYLS	786	67774-32-7
150	DICOFOL	785	115-32-2
151	SELENIUM	779	7782-49-2
152	1,3-BUTADIENE	778	106-99-0
153	1,1,2,2-TETRACHLOROETHANE	776	79-34-5
154	PARATHION	775	56-38-2
155	HEPTACHLORODIBENZO-P-DIOXIN	775	37871-00-4
156	HEXACHLOROCYCLOHEXANE, TECHNICAL GRADE	774	608-73-1
157	TRICHLOROFLUOROETHANE	774	27154-33-2
158	BROMINE	772	7726-95-6
159	AROCLOR 1268	766	11100-14-4
160	PERFLUOROOCTANOIC ACID	761	335-67-1
161	HEPTACHLORODIBENZOFURAN	756	38998-75-3
162	TRIFLURALIN	756	1582-09-8
163	PERFLUOROHEXANESULFONIC ACID	750	355-46-4

164	1,2,3,4,6,7,8,9-OCTACHLORODIBENZOFURAN	744	39001-02-0
165	AMMONIA	742	7664-41-7
166	1,4-DICHLOROBENZENE	727	106-46-7
167	2-METHYLNAPHTHALENE	727	91-57-6
168	2,3,4,7,8-PENTACHLORODIBENZOFURAN	724	57117-31-4
169	NALED	722	300-76-5
170	1,1,2-TRICHLOROETHANE	719	79-00-5
171	1,2-DIPHENYLHYDRAZINE	719	122-66-7
172	1,1-DICHLOROETHANE	718	75-34-3
173	PHORATE	717	298-02-2
174	PERFLUORODECANOIC ACID	712	335-76-2
175	TRICHLOROETHANE	711	25323-89-1
176	ACENAPHTHENE	711	83-32-9
177	TETRACHLOROBIPHENYL	711	26914-33-0
178	PALLADIUM	707	7440-05-3

179	INDENO(1,2,3-CD)PYRENE	705	193-39-5
180	OXYCHLORDANE	705	27304-13-8
181	CRESOL, PARA-	703	106-44-5
182	GAMMA-CHLORDENE	702	56641-38-4
183	TETRACHLOROPHENOL	699	25167-83-3
184	1,2-DICHLOROBENZENE	696	95-50-1
185	1,2-DICHLOROETHENE, TRANS-	691	156-60-5
186	P-XYLENE	687	106-42-3
187	CHLOROETHANE	686	75-00-3
188	ALUMINUM	685	7429-90-5
189	CARBON MONOXIDE	684	630-08-0
190	PHENOL	682	108-95-2
191	CARBON DISULFIDE	681	75-15-0
192	2,4-DIMETHYLPHENOL	679	105-67-9
193	DIBENZOFURAN	676	132-64-9

194	PERFLUORONONANOIC ACID	671	375-95-1
195	PERFLUORODODECANOIC ACID	671	307-55-1
196	ACETONE	671	67-64-1
197	HEXACHLOROETHANE	671	67-72-1
198	BUTYL METHYL PHTHALATE	669	34006-76-3
199	PERFLUOROUNDECANOIC ACID	668	2058-94-8
200	CHLOROMETHANE	664	74-87-3
201	HEXACHLORODIBENZOFURAN	661	55684-94-1
202	HYDROGEN SULFIDE	658	7783-06-4
203	DICHLORVOS	657	62-73-7
204	MONOCHLORODIBENZOFURAN	654	42934-53-2
205	CRESOL, ORTHO-	652	95-48-7
206	HEXACHLORODIBENZO-P-DIOXIN	652	34465-46-8
207	BUTYL BENZYL PHTHALATE	650	85-68-7
208	VANADIUM	650	7440-62-2
209	1,2,4-TRICHLOROBENZENE	647	120-82-1

210	ETHOPROP	645	13194-48-4
211	TETRACHLORODIBENZO-P-DIOXIN	642	41903-57-5
212	BROMOFORM	635	75-25-2
213	PENTACHLORODIBENZOFURAN	632	30402-15-4
214	1,3-DICHLOROBENZENE	628	541-73-1
215	PENTACHLORODIBENZO-P-DIOXIN	626	36088-22-9
216	N-NITROSODIPHENYLAMINE	624	86-30-6
217	2,4-DICHLOROPHENOL	620	120-83-2
218	2,3-DIMETHYLNAPHTHALENE	620	581-40-8
219	2,3,7,8-TETRACHLORODIBENZOFURAN	618	51207-31-9
220	1,4-DIOXANE	617	123-91-1
221	FLUORINE	613	7782-41-4
222	NITRITE	611	14797-65-0
223	DIBROMOCHLOROMETHANE	611	124-48-1
224	CESIUM-137	610	10045-97-3

225	CHROMIC ACID	610	7738-94-5
226	POTASSIUM-40	608	13966-00-2
227	SILVER	608	7440-22-4
228	DINITROTOLUENE	607	25321-14-6
229	FORMALDEHYDE	607	50-00-0
230	COAL TAR PITCH	605	65996-93-2
231	1,2-DICHLOROETHYLENE	605	540-59-0
232	2-BUTANONE	605	78-93-3
233	THORIUM-227	605	15623-47-9
234	ARSENIC ACID	604	7778-39-4
235	NITRATE	604	14797-55-8
236	ANTIMONY	604	7440-36-0
237	ARSENIC TRIOXIDE	604	1327-53-3
238	BENZOPYRENE	603	73467-76-2

239	STROBANE	602	8001-50-1
240	4-AMINOBIPHENYL	602	92-67-1
240	PYRETHRUM	602	8003-34-7
242	ARSINE	602	7784-42-1
242	DIMETHOATE	602	60-51-5
244	BIS(CHLOROMETHYL)ETHER	602	542-88-1
244	CARBOPHENOTHION	602	786-19-6
246	ALPHA-CHLORDENE	601	56534-02-2
246	IODINE-131	601	10043-66-0
246	SODIUM ARSENITE	601	7784-46-5
246	URANIUM-233	601	13968-55-3
250	CRESOLS	598	1319-77-3
251	2,4-D	596	94-75-7
252	DICHLOROBENZENE	596	25321-22-6

253	BUTYLATE	592	2008-41-5
254	2-CHLOROPHENOL	591	95-57-8
255	DIMETHYL FORMAMIDE	585	68-12-2
256	PHENANTHRENE	583	85-01-8
257	DIURON	580	330-54-1
258	4-NITROPHENOL	580	100-02-7
259	TETRACHLOROETHANE	578	25322-20-7
260	DICHLOROETHANE	568	1300-21-6
261	ETHYL ETHER	566	60-29-7
262	DIMETHYLANILINE	563	121-69-7
263	1,3-DICHLOROPROPENE, CIS-	562	10061-01-5
264	ACRYLONITRILE	560	107-13-1
265	1,2,3,4,6,7,8-HEPTACHLORODIBENZO-P-DIOXIN	560	35822-46-9
266	PYRENE	557	129-00-0
267	PHOSPHINE	557	7803-51-2

268	TRICHLOROBENZENE	556	12002-48-1
269	2,6-DINITROTOLUENE	555	606-20-2
270	FLUORIDE ION	550	16984-48-8
271	1,2,3,4,6,7,8-HEPTACHLORODIBENZOFURAN	550	67562-39-4
272	PENTAERYTHRITOL TETRANITRATE	550	78-11-5
273	1,3-DICHLOROPROPENE, TRANS-	549	10061-02-6
274	BIS(2-ETHYLHEXYL)ADIPATE	544	103-23-1
275	CARBAZOLE	544	86-74-8

CAS RN = Chemical Abstracts Service Registry Number

Contact Information

Further information can be obtained by contacting the ATSDR Information Center at:

Agency for Toxic Substances and Disease Registry

Division of Toxicology and Human Health Sciences

1600 Clifton Road NE, Mailstop S106-5

Atlanta, GA 30329

Phone: 1-800-CDC-INFO 888-232-6348 (TTY)

Email: Contact CDC-INFO

ATTENTION ALL Vape users!!

If you vape or even if you don't, these nasty elements are in the vape and still in pollution for all other human beings. They are another example of bad toxins that can harm the body through decades of living. Acetone(Nail polish remover), Acrolein (Weed killer), Formaldehyde, Benzene (Paint stripper) all come out of the body's cells with **ECCRINE POWER!** Adopt now, be happy later!!

Everything! All the elements that are harmful to your body, hundreds of them...All removed with this amazing all-natural, God given, and designed tool!!

Eccrine Power

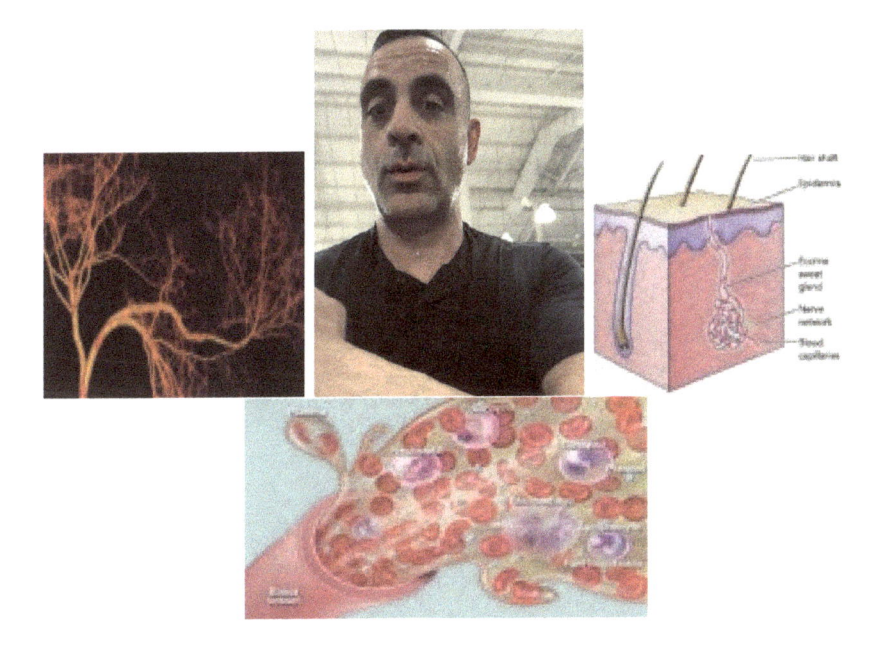

Miscellaneous Fun Facts -

Only horses and humans are equipped with these powerful, pressure built eccrine pores! Again, only horses and humans have them and yeah, both ARE Built to Run & Sweat!

Running is boring, its mental blocks stop most from starting. The start of any run, whether it be one's very first, or a habitual twenty year avid runner, the same thing is true; it's hard to start a run. It's the brain that makes us feel this way. However, what 20+ years has taught me is once you start a run, it's hard to stop. So many times I didn't want to suit up and hit the treadmill, but each time after warming up, I literally went off and ran for over 100 minutes on average. It's all mental. Walk, Run, Run till you're tired, then walk, and JUST REPEAT for success!

71

"A total mishap of theory"

In my prior book, I detailed the facts of the Eccrine pores volume capabilities, which are don't quote me, but upward of 2 liters can come out these pores an hour!! How on earth would the nerve network that doesn't tie into 42 L of water of the body at all, provide that water as opposed to a blood lines which is connected to whole entire pool of 42 L of water in the body?

This is the fundamental gap that they've missed the only thing the nerve network has relevance to is that the bloodlines are also touching them, but it's clear that the water volumes can only come from the bloodlines! Mainly because they are the largest carrier of water in the body, along with cells and the big organs which they are tied to DIRECTLY!' And I mean, directly as evident from real life photos of those organs, like the brain I put in this book that have blood capillaries all on the surface, to capture the water and transfer it to the pores

It's common sense! Really, it's a big mishap theory.

Concluding Quote

"A quote so relevant & dominant to the books content and mission that it most be used as a concluding statement!"

-Giacomo Fasano

"On average, the body of an adult human being contains 60% water. Most of the water in the human body is contained inside our cells. In fact, our billions of cells must have water to live. The total amount of water in our body is found in three main locations: within our cells (two thirds of the water), the space between our cells, and in our blood (one-third of water). For example, a 70-kg man is made up of about 42 L of total water. Actually, the amount of water a body contains varies according to certain contexts: The body of a newborn is composed of more water (75%) than that of an elderly person (50%). Also, the more muscular a body is, the more water the body contains—as body fat has little water. Also, all our vital organs contain different amounts of water: brain, the lungs, the heart, the liver and the kidneys contain a large quantity of water--- between 65 to 85% depending on the organ (2), while bones contain less water (but still 31%!). For all those reasons, water is life."

Once again, its depth and importance evident. However, when combined with the new Eccrine Power connections to blood capillaries, all these inside parts & their water is running directly to the hose of the eccrine pore through these blood lines. This is why, this missing piece of the puzzle cannon be ignored!

Milton Keynes UK
Ingram Content Group UK Ltd.
UKHW022341121024
449428UK00017B/182